YOUR KNOWLEDGE HAS VALUE

AF139914

Bibliographic information published by the German National Library:

The German National Library lists this publication in the National Bibliography; detailed bibliographic data are available on the Internet at http://dnb.dnb.de .

Imprint:

Copyright © 2016 GRIN Verlag, Open Publishing GmbH
Print and binding: Books on Demand GmbH, Norderstedt Germany
ISBN: 9783668526075

This book at GRIN:

http://www.grin.com/en/e-book/373359/finite-difference-methods-with-an-implicit-scheme

Pascal Sturm

Finite difference methods with an implicit scheme

GRIN Publishing

GRIN - Your knowledge has value

Since its foundation in 1998, GRIN has specialized in publishing academic texts by students, college teachers and other academics as e-book and printed book. The website www.grin.com is an ideal platform for presenting term papers, final papers, scientific essays, dissertations and specialist books.

Visit us on the internet:

http://www.grin.com/

http://www.facebook.com/grincom

http://www.twitter.com/grin_com

Implicit Finite Difference Methods

Winter Term 2016/2017

Pascal Sturm

M.Sc. Economics and Finance

Contents

List of Tables

List of Figures

1 Theoretical Aspects of Implicit Finite Difference Methods

Using an explicit scheme for an ap-plication of finite difference methods may lead to stability issues. If one wants to increase the accuracy by raising the number of spatial grid points, the number of time intervals have to be increased to a certain extent in order to sustain a converging behavior. As for quite accurate results ridiculously many grid points in time are needed, the practical use of the explicit scheme is rather limited due to high computational effort. Implicit methods for fi-nite difference methods are designed to overcome these stability limitations imposed by the already mentioned convergence restrictions. Since such methods are unconditionally stable, both accuracy and limited computational effort can be combined.

1.1 Derivation of a Fully Implicit Scheme

As a starting point, assume that the price of an asset follows a geometric Brownian motion:

$$dS_t = (r - q)S_t dt + \sigma S_t dW_t \ . \tag{1}$$

According to Hirsa (2013, p. 115), the value of a derivative on this financial asset specified in (1) must satisfy the Black-Scholes-Merton PDE subject to a given terminal condition at every point in time:

$$\frac{1}{2}\eta^2 S^2 F_{SS}(S,t) + (r-q)SF_S(S,t) - rF(S,t) + F_t(S,t) = 0 \quad \text{s.t. } F(S,T) = f(S,T) \ . \tag{2}$$

In the PDE, η is the so-called diffusion coefficient of the underlying Brownian motion, S the price of the underlying asset, i.e. a stock, r the risk-free interest rate and q the continuously-paid dividend rate.

In order to obtain a numerical solution for the above PDE, the initial variables should be transformed. Furthermore, the continuous pricing problem has to be converted into a discrete one by dividing the feasible time and space domain into a rectangular grid and substituting the partial derivatives with differential quotients. Following this idea, a new set of independent variables is introduced:

$$x = S = ih \quad \forall\, i \in [0,\, M]$$
$$\tau = T - t = nk \quad \forall\, n \in [0,\, N]$$

where i and n are index numbers for the discretized grid points, and h and k are step sizes in space and time direction, respectively. Consequently, the rectangular grid reads:

$$x \in [0,\, x_{\max}] \quad \text{with } x_{\max} = M \cdot h$$
$$\tau \in [0,\, T] \quad \text{with } T = N \cdot k$$

Subsequently, the financial derivative and the terminal condition are also transformed[1]:

$$F(S,t) = u(x,\tau)$$
$$f(S,T) = f(x,0)$$

---

[1]By the definition of τ, the terminal condition becomes an initial condition.

As a final step, the PDE as given in (2) is discretized with the newly introduced variables $x = ih$ and $\tau = nk$ [2]:

$$a(i,n)u_{xx}(i,n) + b(i,n)u_x(i,n) + c(i,n)u(i,n) - u_\tau(i,n) = 0 \quad \text{s.t. } u(x,0) = f(x,0) \qquad (3)$$

By comparing (2) and (3) one may simply define[3]:

$$a(i,n) = a(i) = 0.5\,\eta i^2 h^2 \quad b(i,n) = b(i) = (r-q)ih \quad c(i,n) = c = -r \qquad (4)$$

Since the functional form of the underlying partial derivatives in (3) is not known, they are approximated by differences using Taylor series expansions. For the implementation of the fully implicit scheme, the following approximations around $x = ih$ and $\tau = (n+1)k$ are used:

- $u_{xx}(i, n+1) = \frac{1}{h^2}(U_{i+1}^{n+1} - 2U_i^{n+1} + U_{i-1}^{n+1}) + O(h^2)$

- $u_x(i, n+1) = \frac{1}{2h}(U_{i+1}^{n+1} - U_{i-1}^{n+1}) + O(h^2)$

- $u_\tau(i, n+1) = \frac{1}{k}(U_i^{n+1} - U_i^n) + O(k)$

- $u(i, n+1) = U_i^{n+1} + O(k)$

In contrast to the explicit scheme, the fully implicit scheme employs a backward approximation for the partial time derivative. Plugging these approximations into (3) and dropping the error terms leads to

$$\frac{a}{h^2}\left(U_{i+1}^{n+1} - 2U_i^{n+1} + U_{i-1}^{n+1}\right) + \frac{b}{2h}\left(U_{i+1}^{n+1} - U_{i-1}^{n+1}\right) + cU_i^{n+1} - \frac{1}{k}\left(U_i^{n+1} - U_i^n\right) = 0\,,$$

which can be rearranged for an implicit solution of the values at the next time level:

$$d_1(i, n+1)U_{i-1}^{n+1} + d_2(i, n+1)U_i^{n+1} + d_3(i, n+1)U_{i+1}^{n+1} = d_4(i, n+1) = U_i^n \qquad (5)$$

$$\text{where} \quad d_1(i, n+1) = d_1(i) = -a\frac{k}{h^2} + b\frac{k}{2h} = -\frac{1}{2}\eta^2 i^2 k + \frac{1}{2}(r-q)ik$$

$$d_2(i, n+1) = d_2(i) = a\frac{2k}{h^2} - ck + 1 = 1 + rk + \eta^2 i^2 k$$

$$d_3(i, n+1) = d_3(i) = -a\frac{k}{h^2} - b\frac{k}{2h} = -\frac{1}{2}\eta^2 i^2 k - \frac{1}{2}(r-q)ik$$

In contrast to an explicit scheme, it is not possible to compute U_i^{n+1} directly from the already known values because three unknown values are linked to only one known value. However, since (5) has to hold for all $i = 1, ..., M-1$, a system of $M-1$ equations emerges for the $M+1$ unknown values. Appropriate boundary conditions yield the two missing values for each time step and the terminal conditions give the values in the first time layer. Thus, after adding two additional equations (von-Neumann boundaries) or fixing two values in an appropriate manner (Dirichlet boundary), the $N-1$ linear equation systems can be solved recursively in order to calculate the corresponding derivative price at $\tau = T$.

---

[2]In the functional formulation, the step sizes h and k will be omitted as they are the same for all values. The index numbers i and n determine the underlying dynamics.

[3]Note that all newly introduced coefficients are independent of the time index n.

1.1.1 Exemplary Boundary Conditions for a European Put

The choice of boundary conditions can be crucial for ensuring the accuracy of the pricing algorithm given in (5). This is true as any error on a boundary is propagated through the ongoing finite difference scheme. This makes the entire scheme quite sensitive to the boundary conditions (Hirsa, 2013, p. 121). Hence, a sensible and complete formulation of these conditions at $i = 0$ and $i = M$ is the key to a successful implementation of the underlying pricing problem.

If a European put is the underlying derivative for the recursive pricing algorithm given in (5), one can simply specify the starting point for the implicit scheme by taking the terminal values for a plain-vanilla European put:

$$f(S,0) = u(x,0) = \max(E - S, 0)$$
$$\implies U_i^0 = max(E - ih, 0) \quad \forall \, i \in [0, \, M] \, . \tag{6}$$

For the left-sided boundary, i.e. the stock price is 0, a Dirichlet boundary can be used since it can be assumed that the option value just equals the discounted strike price:

$$U_0^n = Ee^{-r(nk)} \quad \forall \, n \in [1, \, N] \, . \tag{7}$$

Hence for $i = 1$, the equation system in (5) reads

$$d_2(1, n+1)U_1^{n+1} + d_3(1, n+1)U_2^{n+1} = \underbrace{d_4(1, n+1) - d_1(1, n+1)Ee^{-r(n+1)k}}_{d_4^*(1, n+1)} \, . \tag{8}$$

As there is no "natural" bound on the right side of the grid, an artificial one needs to be introduced. It can be assumed that for really high stock prices, the option value does not change if the price of the underlying stock changes slightly, i.e. the Delta of the European put is equal to 0 if $i = M$. Thus, the central approximation for the first derivative around $i = M$ can be used and set equal to zero

$$\frac{\partial U_M^{n+1}}{\partial x} = \frac{1}{2h} \left(U_{M+1}^{n+1} - U_{M-1}^{n+1} \right) \overset{!}{=} 0 \, . \tag{9}$$

Additionally, (5) is also valid at $i = M$

$$d_1(M, n+1)U_{M-1}^{n+1} + d_2(M, n+1)U_M^{n+1} + d_3(M, n+1)U_{M+1}^{n+1} = d_4(M, n+1) \, . \tag{10}$$

Solving (9) for $U_{M+1}^{n+1} = U_{M-1}^{n+1}$ gives the possibility to eliminate U_{M+1}^{n+1} in (10) in order to obtain

$$\underbrace{(d_1(M, n+1) + d_3(M, n+1))}_{d_1^*(M, n+1)} U_{M-1}^{n+1} + d_2(M, n+1)U_M^{n+1} = U_M^n = d_4(M, n+1) \tag{11}$$

The expression in (11) adds an additional equation to the linear system of equations such that there are now M equations for M unknowns at each time layer n [4]. At first glance, it seems that the implicit scheme requires much more computational effort compared to an explicit scheme as a system of equations needs to be solved for every n. However, if the implicit scheme is written in matrix notation, it exhibits a quite nice structure. As shown below, the matrix is tridiagonal which simplifies the solution of the system dramatically.

---

[4]Note that the value U_0^{n+1} is fixed via a Dirichlet boundary and is therefore not an unknown anymore.

1.1.2 The Fully Implicit Scheme in Matrix Notation

The M equations given in (5), (8) and (11) can be written in compact form using matrix notation:

$$\underset{M\times M}{\boldsymbol{A}} \quad \times \quad \underset{M\times 1}{U^{n+1}} \quad = \quad \underset{M\times 1}{d_4} \quad (12)$$

$$
\begin{bmatrix}
d_2(1) & d_3(1) & & & & \\
d_1(2) & d_2(2) & d_3(2) & & \underline{0} & \\
& & \ddots & & & \\
& \underline{0} & & d_1(M-1) & d_2(M-1) & d_3(M-1) \\
& & & & d_1^*(M) & d_2(M)
\end{bmatrix}
\times
\begin{bmatrix}
U_1^{n+1} \\
U_2^{n+1} \\
\vdots \\
U_{M-1}^{n+1} \\
U_M^{n+1}
\end{bmatrix}
=
\begin{bmatrix}
d_4^*(1) \\
d_4(2) \\
\vdots \\
d_4(M-1) \\
d_4(M)
\end{bmatrix}
$$

The straightforward method to solve (12) for U^{n+1} involves taking the inverse of $\boldsymbol{A}$. However, in practice, there are far more efficient solution techniques than matrix inversion. The matrix $\boldsymbol{A}$ has the property that it is tridiagonal, which means that only the diagonal, the super-diagonal and the sub-diagonal elements are non-zero. For such special systems of equations, two efficient methods are outlined in the next Section.

1.2 Solving Tridiagonal System of Linear Equations

In order to solve (12) for U^{n+1}, the tridiagonal structure of the underlying system can be exploited. The first proposed method involves a matrix decomposition into an upper and lower matrix which makes it possible to subdivide the given problem into two minor problems. Assume $\boldsymbol{A} \times U^{n+1} = d_4$ is decomposed into

$$\boldsymbol{A} \times U^{n+1} = (\boldsymbol{L} \times \boldsymbol{R})U^{n+1} = \boldsymbol{L}(\boldsymbol{R} \times U^{n+1}) = \boldsymbol{L} \times y = d_4 \tag{13}$$

This means that the first step employs the so-called Cholesky-decomposition according to $\boldsymbol{A} = \boldsymbol{L} \times \boldsymbol{R}^5$:

$$
\underbrace{
\begin{bmatrix}
d_2 & d_3 & & & \\
d_1 & d_2 & d_3 & & \\
& & \ddots & & \\
& & d_1 & d_2 & d_3 \\
& & & d_1^* & d_2
\end{bmatrix}
}_{\boldsymbol{A}}
=
\underbrace{
\begin{bmatrix}
\alpha_1 & & & & \\
l_1 & \alpha_2 & & & \\
& l_2 & \alpha_3 & & \\
& & \ddots & & \\
& & & l_{M-1} & \alpha_M
\end{bmatrix}
}_{\boldsymbol{L}}
\times
\underbrace{
\begin{bmatrix}
1 & r_1 & & & \\
& 1 & r_2 & & \\
& & \ddots & & \\
& & & 1 & r_{M-1} \\
& & & & 1
\end{bmatrix}
}_{\boldsymbol{R}}
. \tag{14}
$$

From (14) it follows for the first values of α and r

$$d_2(1) = \alpha_1 \qquad \rightarrow \alpha_1 = d_2(1) \tag{15}$$

$$d_3(1) = \alpha_1 r_1 \qquad \rightarrow r_1 = \frac{d_3(1)}{\alpha_1} = \frac{d_3(1)}{d_2(1)}$$

---

[5]In order to keep it well-arranged, the spatial indices are omitted. However, it should be kept in mind that the values for d_1, d_2 and d_3 are different in every row of $\boldsymbol{A}$.

and then in general

$$r_{i-1} = \frac{d_3(i-1)}{\alpha_{i-1}} \qquad i = 2, ..., M$$

$$\alpha_i = d_2(i) - d_1(i)r_{i-1} \quad i = 2, ..., M \qquad (16)$$

$$l_i = d_1(i) \qquad\qquad i = 2, ..., M \quad .$$

Due to the fact that all values in $\boldsymbol{A}$ are time-invariant, this decomposition only needs to be done once as it is the same for every time step[6]. Subsequently, $\boldsymbol{L} \times y = d_4$ can be solved by forward iteration. As d_4 is time-dependent, this step has to be done for every n.

$$\underbrace{\begin{bmatrix} \alpha_1 & & & & \\ l_1 & \alpha_2 & & & \\ & l_2 & \alpha_3 & & \\ & & & \ddots & \\ & & & l_{M-1} & \alpha_M \end{bmatrix}}_{\boldsymbol{L}} \times \underbrace{\begin{bmatrix} y_1 \\ y_2 \\ \vdots \\ y_{M-1} \\ y_M \end{bmatrix}}_{y} = \underbrace{\begin{bmatrix} d_4(1) \\ d_4(2) \\ \vdots \\ d_4(M-1) \\ d_4(M) \end{bmatrix}}_{d_4} \qquad (17)$$

According to (17), it follows directly that

$$\alpha_1 y_1 = d_4(1) \quad \rightarrow y_1 = \frac{d_4(1)}{\alpha_1} \qquad (18)$$

$$y_i = \frac{1}{\alpha_i}(d_4(i) - d_1(i)y_{i-1}) \qquad i = 2, ..., M \qquad (19)$$

which can be used for a backward iterative solution of $\boldsymbol{R} \times U^{n+1} = y$

$$\underbrace{\begin{bmatrix} 1 & r_1 & & & \\ & 1 & r_2 & & \\ & & \ddots & & \\ & & & 1 & r_{M-1} \\ & & & & 1 \end{bmatrix}}_{\boldsymbol{R}} \times \underbrace{\begin{bmatrix} U_1^{n+1} \\ U_2^{n+1} \\ \vdots \\ U_{M-1}^{n+1} \\ U_M^{n+1} \end{bmatrix}}_{U^{n+1}} = \underbrace{\begin{bmatrix} y_1 \\ y_2 \\ \vdots \\ y_{M-1} \\ y_M \end{bmatrix}}_{y} . \qquad (20)$$

Using (20), one can finally solve for the unknown values U^{n+1}

$$U_M^{n+1} = y_M \qquad (21)$$

$$U_i^{n+1} = y_i - r_i U_{i+1}^{n+1} = y_i - \frac{d_3(i)}{\alpha(i)}U_{i+1}^{n+1} \qquad i = M-1, ..., 1 \qquad (22)$$

Although this workaround seems quite time-consuming, solving a tridiogonal system of equation with the above algorithm requires only a little bit more computational effort compared to an explicit scheme, to be exact twice as many operations per time step (Wilmott et al., 1995, p. 144). Yet it can be shown that the implicit scheme is unconditionally stable which means that it is not necessary to increase N to really large numbers for a finer spatial grid[7]. This more than compensates the loss in time for solving the above linear system. Thus, implicit methods can be considered superior to explicit methods.

---

[6]It should be noted that actually only (15) and (16) are needed for later steps. Hence, only these values must be stored somewhere.

[7]A proof of the unconditional stability of the implicit scheme is given by Brandimarte (2006, p. 311).

For the sake of completeness, an algorithm for a Gauss elimination is given as a second method in the following. As for the first method, there are some calculations which do not involve d_4 and are therefore time-invariant:

$$d_1^*(i) := \frac{d_1(i)}{d_2(i-1)} \qquad\qquad i = 2, ..., M \qquad\qquad (23)$$

$$d_2^*(i) := d_2(i) - d_1^*(i)d_3(i-1) \qquad i = 2, ..., M \ . \qquad\qquad (24)$$

As a next step, one needs to do a forward iteration according to

$$d_4^*(i) := d_4(i) - d_1^*(i)d_4(i-1) \qquad i = 2, ..., M \qquad\qquad (25)$$

and finally a backward iteration which results in

$$U_M^{n+1} = \frac{d_4^*(M)}{d_2(M)} \qquad\qquad\qquad\qquad (26)$$

$$U_i^{n+1} = \frac{d_4^*(i) - d_3(i)U_{i+1}^{n+1}}{d_2^*(i)} \qquad i = M-1, ..., 1 \ . \qquad\qquad (27)$$

As for the first method, the forward and backward iterations have to be done on each time layer n. The implementation of both methods using MATLAB and a comparison with respect to computational time are given below.

1.3 Derivation of the Crank-Nicolson Scheme

So far, the explicit and the implicit scheme were derived using a forward or a backward approximation in time, respectively. As a central approximation for the first derivative in space was proven to be more accurate than a forward or backward iteration, one could follow the same line of argumentation and simply combine them to a central approximation with respect to time. However, such approximations with step size k in each direction usually result in very unstable solutions which make them defective for practical use (Wilmott et al., 1995, p. 138).

This gave rise to the Crank-Nicolson (CN) method by Crank & Nicolson (1947) who managed to combine a forward and backward approximation in a feasible way in order to make use of a more accurate central approximation in time. The basic idea is to consider a grid point which is actually not on the underlying grid but positioned halfway between two time layers[8]. This means that (2) is discretized around $x = ih$ and $\tau = (n + \frac{1}{2})k$. For a discretization around this point, there is no other possibility than using a central approximation since a forward or backward approximation would involve a differential quotient with a grid point outside the actual grid. For a derivation of this central approximation, one needs to consider a Taylor series expansion in time direction using an appropriate time step. Obviously the appropriate time step is $\frac{k}{2}$:

$$u\left(x, \tau \pm \frac{k}{2}\right) = u(x, \tau) \pm \frac{k}{2}u_\tau + \frac{1}{2}\left(\frac{k}{2}\right)^2 u_{\tau\tau} \pm \frac{1}{6}\left(\frac{k}{2}\right)^3 u_{\tau\tau\tau} + \dots \ . \qquad (28)$$

---

[8]For the original CN method the grid point is halfway between n and $n+1$. However, there are also other methods which vary the positioning between two time layers.

Subtracting $u\left(x, \tau - \frac{k}{2}\right)$ from $u\left(x, \tau + \frac{k}{2}\right)$ leads to

$$u\left(x, \tau + \frac{k}{2}\right) - u\left(x, \tau - \frac{k}{2}\right) = ku_\tau + \frac{1}{3}\left(\frac{k}{2}\right)^3 u_{\tau\tau\tau} + O\left(k^5\right)$$

which can finally be solved for

$$u_\tau = \frac{u(x, \tau + \frac{k}{2}) - u(x, \tau - \frac{k}{2})}{k} + O\left(k^2\right)$$

and is in fact the same as

$$u_\tau\left(i, n + \frac{1}{2}\right) = \frac{U_i^{n+1} - U_i^n}{k} + O\left(k^2\right) \tag{29}$$

when a discretization around $x = ih$ and $\tau = (n + \frac{1}{2})k$ is considered. As this imaginary point is exactly halfway off the grid between the points which were used for the derivation of the explicit and the fully implicit scheme, the partial derivatives with respect to x can be approximated by an average of the differential quotients which are used for the implementation of the explicit and fully implicit method (Hirsa, 2013, p. 120):

$$u_{xx}\left(i, n + \frac{1}{2}\right) = \frac{1}{2h^2}\left(U_{i+1}^{n+1} - 2U_i^{n+1} + U_{i-1}^{n+1} + U_{i+1}^n - 2U_i^n + U_{i-1}^n\right) + O\left(h^2 + \frac{k^2}{4}\right) \tag{30}$$

$$u_x\left(i, n + \frac{1}{2}\right) = \frac{1}{4h}\left(U_{i+1}^{n+1} - U_{i-1}^{n+1} + U_{i+1}^n - U_{i-1}^n\right) + O\left(h^2 + \frac{k^2}{4}\right) \tag{31}$$

By doing so, an additional error term of order k^2 is introduced for u_{xx} and u_x. Using appropriate Taylor series expansions according to $u\left(x \pm h, \tau \pm \frac{k}{2}\right)$, the same formulas can be derived. Additionally, it can be shown that all terms of the order k cancel out.

Finally, adding $u\left(x, \tau + \frac{k}{2}\right)$ and $u\left(x, \tau - \frac{k}{2}\right)$ from (28) leads to

$$u\left(i, n + \frac{1}{2}\right) = \frac{1}{2}\left(U_i^{n+1} + U_i^n\right) + O\left(\frac{k^2}{4}\right). \tag{32}$$

Now, (29), (30), (31) and (32) can be plugged into (3). After some cancellations and rearrangements, the following expression is obtained

$$d_1\left(i, n + \frac{1}{2}\right)U_{i-1}^{n+1} + d_2\left(i, n + \frac{1}{2}\right)U_i^{n+1} + d_3\left(i, n + \frac{1}{2}\right)U_{i+1}^{n+1} = d_4\left(i, n + \frac{1}{2}\right) \tag{33}$$

where

$$d_1\left(i, n + \frac{1}{2}\right) = d_1(i) = -a\frac{k}{2h^2} + b\frac{k}{4h} = -\frac{1}{4}\eta^2 i^2 k + \frac{1}{4}(r - q)ik$$

$$d_2\left(i, n + \frac{1}{2}\right) = d_2(i) = 1 + a\frac{k}{h^2} - c\frac{k}{2} = 1 + \frac{1}{2}\eta^2 i^2 k + \frac{1}{2}rk$$

$$d_3\left(i, n + \frac{1}{2}\right) = d_3(i) = -a\frac{k}{2h^2} - b\frac{k}{4h} = -\frac{1}{4}\eta^2 i^2 k - \frac{1}{4}(r - q)ik$$

$$d_4\left(i, n + \frac{1}{2}\right) = -d_1\left(i, n + \frac{1}{2}\right)U_{i-1}^n + \left(2 - d_2\left(i, n + \frac{1}{2}\right)\right)U_i^n - d_3\left(i, n + \frac{1}{2}\right)U_{i+1}^n.$$

As it is the case for the explicit and implicit scheme, d_1, d_2 and d_3 are independent of time. Moreover, it can be noted that the functional expression in (33) is also simply something like a mix of the explicit and the fully implicit scheme. The right-hand side, d_4, can be calculated explicitly since all values in n are already known. On the contrary, the values at time layer $n+1$ are again only implicitly known to us. As in the case of the fully implicit scheme, (33) holds for $i = 1, ..., M-1$ which means that $M-1$ equations can be used to solve for $M+1$ unknowns. Hence, the system can only be solved after specifying proper boundary conditions.

1.3.1 Exemplary Boundary Conditions for a European Put

For a European put, the boundaries can be derived in a similar way as it is done above for the fully implicit scheme. The terminal values again serve as starting values for the recursive calculation, i.e.

$$f(S,0) = u(x,0) = \max(E - S, 0)$$
$$\Longrightarrow U_i^0 = max(E - ih, 0) \quad \forall\, i \in [0,\, M]\,. \tag{34}$$

The argumentation for the the left-sided boundary ($i = 0$) also stays the same, meaning that all values can be set equal to the discounted strike:

$$U_0^n = Ee^{-r(nk)} \quad \forall\, n \in [1,\, N]\,. \tag{35}$$

Consequently, the equation system in (33) for $i = 1$ reads

$$d_2\,(1)\,U_1^{n+1} + d_3\,(1)\,U_2^{n+1} = \underbrace{d_4\left(1, n + \frac{1}{2}\right) - d_1\,(1)\,Ee^{-r(n+1)k}}_{d_4^*\left(1, n+\frac{1}{2}\right)} \tag{36}$$

$$\text{with} \quad d_4\left(1, n + \frac{1}{2}\right) = d_1\,(1)\,Ee^{-r(nk)} + (2 - d_2\,(1))\,U_1^n - d_3\,(1)\,U_2^n\,.$$

Lastly, for really high stock prices, i.e. $i = M$, the assumption that the Delta for a European put equals zero is made again. Hence it is straightforward to use

$$\frac{\partial U_M^{n+1}}{\partial x} = \frac{1}{2h}\left(U_{M+1}^{n+1} - U_{M-1}^{n+1}\right) \overset{!}{=} 0$$
$$\text{and} \quad \frac{\partial U_M^n}{\partial x} = \frac{1}{2h}\left(U_{M+1}^n - U_{M-1}^n\right) \overset{!}{=} 0$$

in order to establish that $U_{M+1}^{n+1} = U_{M-1}^{n+1}$ and $U_{M+1}^n = U_{M-1}^n$. These expressions can be used to eliminate the emerging ghost points when (33) is evaluated at $i = M$. Hence, the required M-th equation for the system of equation is given by

$$\underbrace{(d_1(M) + d_3(M))}_{d_1^*(M)}U_{M-1}^{n+1} + d_2(M)U_M^{n+1} = \underbrace{\underbrace{-(d_1(M) + d_3(M))}_{-d_1^*(M)}U_{M-1}^n + (2 - d_2(M)U_M^n}_{d_4^*\left(M, n+\frac{1}{2}\right)}\,. \tag{37}$$

Using (33), (35), (36) and (37), a European put price can be calculated according to the CN scheme. As for the fully implicit scheme, the system of equations is tridiagonal.

1.3.2 The Crank-Nicolson Scheme in Matrix Notation

As before, the M equations (33), (36) and (37) can be written in matrix notation.

$$\underset{M \times M}{A} \quad \times \quad \underset{M \times 1}{U^{n+1}} \quad = \quad \underset{M \times 1}{d_4} \quad (38)$$

$$\begin{bmatrix} d_2(2) & d_3(2) & & & & \\ d_1(3) & d_2(3) & d_3(3) & & \underline{0} & \\ & & & \ddots & & \\ & \underline{0} & & d_1(M-1) & d_2(M-1) & d_3(M-1) \\ & & & & d_1^*(M) & d_2(M) \end{bmatrix} \times \begin{bmatrix} U_1^{n+1} \\ U_2^{n+1} \\ \vdots \\ U_{M-1}^{n+1} \\ U_M^{n+1} \end{bmatrix} = \begin{bmatrix} d_4^*(1) \\ d_4(2) \\ \vdots \\ d_4(M-1) \\ d_4^*(M) \end{bmatrix}$$

with

$$\underset{M \times 1}{d_4} \quad = \quad \underset{M \times M}{B} \quad \times \quad \underset{M \times 1}{U^n} \quad + \quad \underset{M \times 1}{f} \quad (39)$$

$$\begin{bmatrix} d_4^*(1) \\ d_4(2) \\ \vdots \\ d_4(M-1) \\ d_4^*(M) \end{bmatrix} = \begin{bmatrix} 2 - d_2 & -d_3 & & & \\ -d_1 & 2 - d_2 & -d_3 & \underline{0} & \\ & & \ddots & & \\ & \underline{0} & -d_1 & 2 - d_2 & -d_3 \\ & & & -d_1^* & 2 - d_2 \end{bmatrix} \times \begin{bmatrix} U_1^n \\ U_2^n \\ \vdots \\ U_{M-1}^n \\ U_M^n \end{bmatrix} + \begin{bmatrix} -d_1(1)(U_0^{n+1} + U_0^n) \\ 0 \\ \vdots \\ \vdots \\ 0 \end{bmatrix}$$

Hence, solving this system of equations is not different from solving the equation for the implicit scheme given in (12). This is true since everything on the right-hand side of (38) can be evaluated explicitly using the known U_i^n. Consequently, the problem reduces to first calculating the right-hand side according to (39) and then solving (38).

Just as for the implicit scheme, convergence of the CN method does not require particularly small time steps. As shown and explained below, the CN method typically converges faster than the implicit scheme and thus gives more precise approximations of unknown solutions. However, although the CN method is proven to be unconditionally stable, relatively large values for k compared to h may negatively affect the actual performance of the scheme due to spurious oscillations in the numerical solution (Fusai & Roncoroni, 2008, p. 107). These spurious oscillations are introduced by a kink in the initial condition and only disappear really slowly if N is not sufficiently increased[9]. Nevertheless, the CN method does not lead to such severe stability issues for which explicit methods are known for.

---

[9]In the Appendix, examples for spurious oscillation can be found.

2 Numerical Application for a European Put

2.1 Results and Interpretations

For a practical implementation of the implicit scheme as well as the CN scheme, the following data is used:

$$x = S = 200 \quad E = 200 \quad T = 1.00 \quad r = 0.04 \quad q = 0.06 \quad \eta = 0.25 \ . \tag{40}$$

With respect to the rectangular grid, an upper bound of $S_{Max} = x_{Max} = 1200$ is chosen and the spatial domain is divided up into $M = 120$ space intervals. The time dimension is sliced into $N = 100$ intervals implying step sizes of $h = 10$ in space and of $k = 0.01$ in time. Subsequently, (12) and (38) can be used in order to obtain numerical approximations of the exact BSM solution for a European put for the fully implicit method and the CN method, respectively. While the fair BSM price of a European put option with the above characteristics amounts up to 20.8886, the numerical solutions yield only 20.7655 (-0.59 %) in case of the fully implicit scheme and 20.7951 (-0.45 %) if the CN scheme is used. Apparently, both schemes exhibit the same characteristic also features by the explicit scheme, namely an underestimation of the true value when the strike of the underlying option coincides with a spatial grid point. Thus, the absolute errors of 0.1231 (implicit scheme) and 0.0936 (CN scheme) for $S = 200$ are also the maximum absolute errors which can be seen in Figure 1.

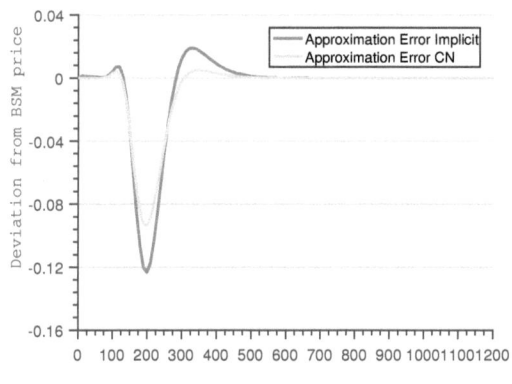

Figure 1: Errors of numerical solutions $(U - u^{BSM}(S, \tau = T))$ depending on S

As it is the case for the explicit scheme, the resulting errors for full time to maturity are introduced at the very first approximation step in $n = 1$. This is illustrated in Figure 2. The payoff function for the European put which is used as the initial condition for both schemes is non-differentiable at the respective strike. The substitution of the partial derivatives in the BSM PDE by differences obtained from Taylor series expansions at this value introduces an approximation error which is simply propagated through the underlying grid. Due to the fact that both methods are consistent, the errors diminish with ongoing time steps[10]. Nevertheless, the errors also spread in space, giving rise to the same wavy behavior as observed for the explicit scheme.

---

[10]The proof of consistency is provided further below.

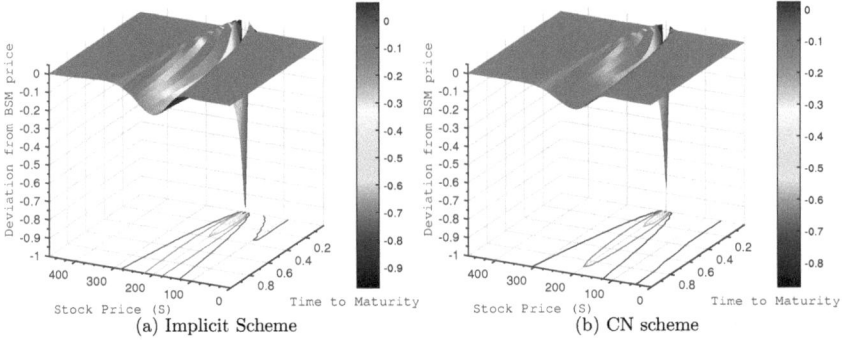

(a) Implicit Scheme (b) CN scheme

Figure 2: 3-D Plot for Error of Numerical Solutions depending on S and τ

Comparing both error evolutions a little bit more closely, some additional observations can be made. Firstly, the error which is introduced in $n = 1$ is smaller for the CN scheme. Secondly, it seems that the converging behavior, i.e. the error reduction per time step, of the CN scheme is a bit faster compared to the fully implicit scheme. This can be verified by inspecting the underlying contours of both plots. In general, the CN scheme displays a smoother convergence than the implicit scheme. Certainly, a combination of these two observations can explain why the approximation errors for full time to maturity in Figure 1 are smaller for the CN scheme. The smoother convergence or figuratively speaking the fact that the wave fades out more smoothly for the CN scheme, might also explain why the feedback effect of the wave, i.e. the positive approximation errors around $S = 300$, is much stronger for the implicit scheme. Thirdly, it initially seems that for all points in time, the error stabilizes around zero for the given boundaries in space. However, looking at Figure 2b, a strange contour line is visible around $S = 50$ for full time to maturity. An explanation for this contour line can be found by zooming in Figure 1 for small values of S. As illustrated in Figure 3, there seems to be a little problem with the Dirichlet boundary for the implicit scheme. So far, there is no sensible explanation for this anomaly, especially as the CN scheme which uses the same reasoning and computational algorithm does not encounter the same problem.

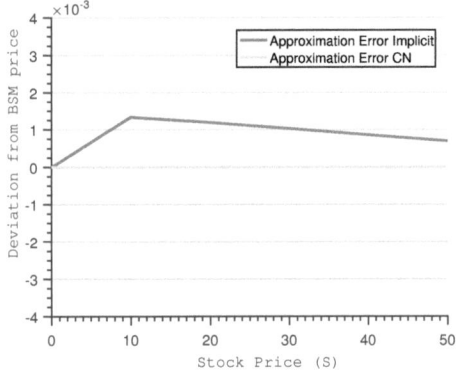

Figure 3: Errors of numerical solutions $(U - u^{BSM}(S, \tau = T))$ depending on S (zoom)

11

Before Section 3 explains theoretically why the CN scheme shows a faster convergence and hence yields better numerical results, a brief discussion about the implementation in MAT-LAB can be found below. Additionally, a short efficiency analysis with respect to computational time is given.

2.2 Technical Implementation

The technical implementation of this Section can be found in *Task1.m* and *Task2.m*. These two main scripts make use of several functions for the fully implicit method as well as for the CN method. For the fully implicit method, three functions are available. *Implicit_put.m* defines the coefficients according to (5) and corrects for the von-Neumann boundary condition as given in (11). Then a Cholesky decomposition is employed by the built-in MATLAB function *lu*. This decomposition only involves the coefficients d_1, d_2 and d_3 which are time-invariant. Hence, this decomposition only has to be done once. After the Dirichlet boundary with respect to (8) is implemented, the matrices resulting from the Cholesky decomposition are used to solve the linear system in (12) by matrix inversion at every time step. This is done by making us of the MATLAB operator "\". However, this procedure is quite inefficient as the inversion and matrix multiplication involves many calculations that are actually not needed, like multiplications with 0. This is reflected in a computation time of 0.0337 seconds[11].

Implicit_put2.m works in principle as *Implicit_put.m* but uses the algorithm given in (13) - (22) for an efficient decomposition of A and calculation of U^{n+1}. Compared to *Implicit_put.m*, only necessary arithmetical operations are executed which leads to a dramatically lower computational time of 0.0033 seconds. Hence, this function works 10 times faster as the first version. *Implicit_put3.m* employs the Gauss algorithm given in (23) - (27) for the solution of the system of equation. This results in a run time of 0.0034 seconds which is in fact the same as for *Implicit_put2.m*. This is not surprising as approximately the same number of operations are used. Nevertheless, *Implicit_put2.m* is chosen for all further calculations of the fully implicit scheme.

The implementation of the CN scheme is compared in four functions. All functions are basically programmed to solve (38). In general, *CN_put.m*, *CN_put2.m* and *CN_put3.m* are comparable to the respective functions for the implicit scheme. The only difference is the calculation of d_4 at every time step. In all three functions, this is done by matrix calculation using the matrix B as outlined in (39). This matrix consists of time-invariant coefficients and only needs to be set up once. However, this matrix operation involves an increasing amount of multiplications with 0 for high values of M and is thus inefficient. For this reason, *CN_put4.m* calculates d_4 only using all necessary values while sticking to the efficient LR-decomposition as implemented in *CN_put2.m*. The computational effort for all 4 functions amounts up to 0.0345, 0.0053, 0.0058 and 0.0045 seconds respectively. These values reflect the previous argumentation about performance issues. Based on these numbers, *CN_put4.m* is chosen for further calculations with the CN scheme.

Comparing the fully implicit scheme with the CN scheme in their most efficient forms, the CN scheme needs 1.36 times more seconds than the fully implicit scheme but also reduces the maximum absolute error to 76,04 %. Hence, there is a tradeoff between accuracy and computational time, switching between both schemes. This is investigated more closely below.

---

[11]Of course, the absolute run times can vary for different machines. Nevertheless, the relative differences contain information about the relative performance.

3   Truncation Error and Rate of Convergence

The numerical errors observed in the Section above may arise from various sources. One major source can be attributed to the truncation error which stems from the difference approximation of the differential operators (Kwok, 2008, p. 337). Corresponding to Lax's Equivalence Theorem, the local truncation error of a numerical scheme must necessarily tend to zero for $h, k \to 0$ if the numerical solution is supposed to converge to the true value. This property is called consistency which ensures convergence in the limit. Since step sizes really close to zero normally coincide with enormous computational effort in numerical applications, it is also of major interest how fast the numerically obtained values converge to the true value.

According to Kwok (2008, p. 332), the order of accuracy or rate of convergence of a scheme is defined to be the order in powers of h and k in the leading truncation error terms. These leading error terms provide an **asymptotic** measure for the accuracy of a numerical solution[12]. In general, if the leading truncation error is $O(h^m, k^j)$, then the numerical scheme is said to be $m - th$-order space accurate and $j - th$-order time accurate. As outlined by Kwok (2008, p. 332), the numerical solution of a $j - th$-order time accurate scheme is reduced by a factor of $\frac{1}{2^j}$ when the time step is reduced by $\frac{1}{2}$, i.e. twice as many grid points in time as before. A same argumentation should also hold for the order of space accuracy.

Following this line of argumentation, a comparison of the respective truncation errors might explain the difference between the fully implicit scheme and the CN scheme which is observed in Figure 1. The common way of calculating the truncation error of a finite difference solution is to expand $u(x, \tau)$ around the point at which it is discretized and then insert these Taylor series expressions into discretized PDE. After collecting terms and simplifying the whole expression, one receives the resulting truncation error of an underlying scheme. Following this recipe, the derivation of the truncation error for the fully implicit scheme employs Taylor series expansions around $x = ih$ and $\tau = (n + 1)k$

$$u(x \pm h, \tau) = u(x, \tau) \pm h u_x + \frac{1}{2} h^2 u_{xx} \pm \frac{1}{6} h^3 u_{xxx} + \frac{1}{24} h^4 u_{xxxx} \pm \dots$$

$$u(x, \tau - k) = u(x, \tau) - k u_\tau + \frac{1}{2} k^2 u_{\tau\tau} - \dots$$

which are plugged into the discretized PDE

$$\frac{a}{h^2} \left(U_{i+1}^{n+1} - 2U_i^{n+1} + U_{i-1}^{n+1} \right) + \frac{b}{2h} \left(U_{i+1}^{n+1} - U_{i-1}^{n+1} \right) + cU_i^{n+1} - \frac{1}{k} \left(U_i^{n+1} - U_i^n \right) = 0$$

in order to obtain

$$a u_{xx} + \frac{a}{12} h^2 u_{xxxx} + b u_x + b \frac{h^2}{6} u_{xxx} + cu - u_\tau + \frac{1}{2} k u_{\tau\tau} + O(h^4, k^2) = 0 .$$

Due to the fact that $a(i, n) u_{xx}(i, n) + b(i, n) u_x(i, n) + c(i, n) u(i, n) - u_\tau(i, n) = 0$, there remains the truncation error according to

$$T^{Imp}(i, n) = h^2 \left(\frac{a}{12} u_{xxxx} + \frac{b}{6} u_{xxx} \right) + k \left(\frac{1}{2} u_{\tau\tau} \right) + O(h^4, k^2)$$

$$T^{Imp}(i, n) = O(h^2, k) .$$

Hence the fully implicit scheme is first-order time accurate and second-order space accurate.

---

[12]A quite basic illustration in the Appendix tries to explain this with some plots.

A similar analysis can be carried out for the CN scheme. This time, one needs to use appropriate Taylor series expansions around $x = ih$ and $\tau = (n + \frac{1}{2})k$. The corresponding Taylor series expansions are:

$$u(x \pm h, \tau + \frac{k}{2}) = u(x, \tau) \pm hu_x + \frac{1}{2}h^2 u_{xx} \pm \frac{1}{6}h^3 u_{xxx} + \frac{1}{24}h^4 u_{xxxx} + \frac{k}{2}u_\tau + \frac{k^2}{8}u_{\tau\tau} + \frac{k^3}{48}u_{\tau\tau\tau}$$
$$\pm h\frac{k}{2}u_{x\tau} + \frac{1}{2}h^2\frac{k}{2}u_{xx\tau} \pm \frac{1}{2}h\frac{k^2}{4}u_{x\tau\tau} + \frac{1}{4}h^2\frac{k^2}{4}u_{xx\tau\tau} + \ldots$$

$$u(x \pm h, \tau - \frac{k}{2}) = u(x, \tau) \pm hu_x + \frac{1}{2}h^2 u_{xx} \pm \frac{1}{6}h^3 u_{xxx} + \frac{1}{24}h^4 u_{xxxx} - \frac{k}{2}u_\tau + \frac{k^2}{8}u_{\tau\tau} - \frac{k^3}{48}u_{\tau\tau\tau}$$
$$\mp h\frac{k}{2}u_{x\tau} - \frac{1}{2}h^2\frac{k}{2}u_{xx\tau} \pm \frac{1}{2}h\frac{k^2}{4}u_{x\tau\tau} + \frac{1}{4}h^2\frac{k^2}{4}u_{xx\tau\tau} + \ldots$$

$$u(x, \tau + \frac{k}{2}) = u(x, \tau) \pm \frac{k}{2}u_\tau + \frac{k^2}{8}u_{\tau\tau} \pm \frac{k^3}{48}u_{\tau\tau\tau} + \ldots \quad .$$

Plugging them into the discretized PDE

$$\frac{a}{2h^2}\left(U_{i+1}^{n+1} - 2U_i^{n+1} + U_{i-1}^{n+1} + U_{i+1}^n - 2U_i^n + U_{i-1}^n\right) + \frac{b}{4h}\left(U_{i+1}^{n+1} - U_{i-1}^{n+1} + U_{i+1}^n - U_{i-1}^n\right)$$
$$+ \frac{c}{2}\left(U_i^{n+1} + U_i^n\right) - \frac{1}{k}\left(U_i^{n+1} - U_i^n n\right) = 0$$

yields

$$\frac{a}{2h^2}\left(2h^2 u_{xx} + \frac{1}{6}h^4 u_{xxxx} + h^2\frac{k^2}{4}u_{xx\tau\tau}\right) + \frac{b}{4h}\left(4hu_x + \frac{2}{3}h^3 u_{xxx} + 2h\frac{k^2}{4}u_{x\tau\tau}\right)$$
$$+ \frac{c}{2}\left(2u + \frac{k^2}{4}u_{\tau\tau}\right) - \frac{1}{k}\left(ku_\tau + \frac{1}{3}\frac{k^3}{8}u_{\tau\tau\tau}\right) + O(h^3 + k^3) = 0$$

which can be simplified to

$$\underbrace{au_{xx} + bu_x + cu - u_\tau}_{=0} + h^2\left(\frac{a}{12}u_{xxxx} + \frac{b}{6}u_{xxx}\right) + k^2\left(\frac{a}{8}u_{xx\tau\tau} + \frac{b}{8}u_{x\tau\tau} + \frac{c}{8}u_{\tau\tau} - \frac{1}{24}u_{\tau\tau\tau}\right) = 0$$

$$\Longrightarrow T^{CN}\left(i, n + \frac{1}{2}\right) = O(h^2, k^2) .$$

Thus, the CN scheme exhibits second-order temporal and spatial accuracy. Consequently, the CN scheme should converge faster than the fully implicit scheme as its truncation error is of higher order with respect to time. This could be the reason for the more accurate results illustrated in Figure 1.

As a next step, the truncation error can be used to estimate how many more additional grid points are necessary for a given error reduction. In order to do so, consider the truncation error for the implicit scheme as given above:

$$T_M^N(i, n) = h^2\underbrace{\left(\frac{a}{12}u_{xxxx} + \frac{b}{6}u_{xxx}\right)}_{=:v \text{ and independent of } h} + k\underbrace{\left(\frac{1}{2}u_{\tau\tau}\right)}_{=:w \text{ and independent of } k} .$$

If the underlying grid is refined with $4N$ and $2M$, one could expect

$$T_{2M}^{4N}(i, n) = \left(\frac{h}{2}\right)^2 v + \frac{k}{4} w \approx \frac{1}{4} T_M^N(i, n)$$

which means that the error shrinks down to $\frac{1}{4}$ if 8 times more grid points are used. For the underlying numerical application in Section 2, a maximum absolute of 0.1231 is obtained for $N = 100$ and $M = 120$. An increase to $N = 400$ and $M = 240$ indeed leads to an error of 0.0306 which is approximately 25 % of the previous error. This tradeoff between additional time and increased accuracy seems quite promising, but if one wants to reduce the error to 1%, one needs to use 1,000 times as many grid points:

$$\frac{1}{100} T_M^N(i, n) \approx \left(\frac{h}{10}\right)^2 v + \frac{k}{100} w = T_{10M}^{100N}(i, n)$$

Numerically, this can be confirmed as the error for $S = 200$ only amounts to 0.0012 after M is increased to 1,200 and N to 10,000. This requires a relatively large increase in computational effort, i.e. from 0.0033 seconds to 1.8814 seconds (multiplied approximately by 570).

For the CN scheme, the original approximation for $N = 100$ and $M = 120$ is equal to 0.0936. As the CN is of second-order temporal accuracy, only twice as many grid points in time and space are needed for an error reduction to 25 %. In order to decrease the error to 1 %, the higher-order accuracy only requires 10 times more grid points in time, i.e. only 100 times more grid points in total. The error reduction to $9.3172 * 10^{-4}$ goes along with an increase in computational time from 0.0045 to only 0.2663 (multiplied approximately by 59).

Thus, it seems to be a more effective way to enhance the accuracy by increasing the leading error terms in the truncation error. In the literature, this idea was firstly addressed by Richardson (1911). He introduced a method of extrapolation which, in its most basic form, seeks to combine two grids with different step sizes in such a way that leading error terms of the truncation error cancel out. In the following, suitable grid combinations for the fully implicit scheme as well as for the CN scheme are developed that aim at increasing accuracy with less computational effort.

3.1 Richardson Extrapolation for the Fully Implicit Scheme

Generally, the truncation error is implicitly given by the difference between the exact continuous solution $u(x, \tau)$ and the discretized solution U_M^N, using N time intervals and M space intervals. Thus, the following relation can be established:

$$U_M^N = u(x, \tau) + T_M^N .$$

Considering the truncation error of the implicit scheme $T_M^N = h^2 v + kw$, the following numerical solutions for the implicit scheme can be obtained:

$$U_M^N = u(x, \tau) + h^2 v_2 + h^4 v_4 + h^6 v_6 + kw_1 + k^2 w_2 + k^3 w_3 + O(h^8 + k^4)$$
$$U_{2M}^{2N} = u(x, \tau) + \frac{1}{4} h^2 v_2 + \frac{1}{16} h^4 v_4 + \frac{1}{64} h^6 v_6 + \frac{1}{2} kw_1 + \frac{1}{4} k^2 w_2 + \frac{1}{8} k^3 w_3 + O(h^8 + k^4)$$
$$U_{2M}^{4N} = u(x, \tau) + \frac{1}{4} h^2 v_2 + \frac{1}{16} h^4 v_4 + \frac{1}{64} h^6 v_6 + \frac{1}{4} kw_1 + \frac{1}{16} k^2 w_2 + \frac{1}{64} k^3 w_3 + O(h^8 + k^4)$$

In a next step, these three different grids can be combined in a suitable way in order to eliminate higher order terms resulting in a better speed of convergence. For instance, consider the following combinations:

$$\text{1)} \quad \frac{1}{3}\left(4U_{2M}^{2N} - U_M^N\right) = u(x,\tau) + O\left(h^4 + k\right)$$

$$\text{2)} \quad 2U_{2M}^{2N} - U_M^N = u(x,\tau) + O\left(h^2 + k^2\right)$$

$$\text{3)} \quad \frac{1}{3}\left(4U_{2M}^{4N} - U_M^N\right) = u(x,\tau) + O\left(h^4 + k^2\right)$$

Obviously, the third combination seems to be the best choice as it eliminates the next higher-order term in time as well as in space. Nevertheless, this comes at the cost of more computational effort, i.e. an additional grid with 8 times more grid points have to be calculated. Still, the relation between gain in accuracy and loss in computational time is optimized by choosing the third combination. This can be numerically as well as graphically verified by looking at Table 1 and Figure 4.

Table 1: Results for Richardson Extrapolation Implicit Scheme

	approx. error	comp. time	$\dfrac{\text{approx. error}}{\text{approx. error of } U_M^N}$	$\dfrac{\text{comp. time}}{\text{comp. time of } U_M^N}$
U_M^N	0.1231	0.0033	100 %	100 %
$\frac{1}{3}\left(4U_{2M}^{2N} - U_M^N\right)$	0.0095	0.0123	7.72 %	361.76 %
$2U_{2M}^{2N} - U_M^N$	0.0474	0.0114	38.51 %	345.45 %
$\frac{1}{3}\left(4U_{2M}^{4N} - U_M^N\right)$	0.000268	0.0194	0.22 %	587,88 %

In general, the respective orders of accuracy for all three methods are in line with the observed approximation errors in Figure 4. For a comparison of the development of the approximation error with ongoing time steps, Figure 5 opposes the error development for the standard implicit scheme with the Richardson extrapolation using the third combination. It is not surprising that the error for the Richardson extrapolation is much smaller in absolute terms at $n = 1$ as well as that this error fades out much faster. This is practically the visualization of how the rate of convergence influences a numerical solution in finite difference methods.

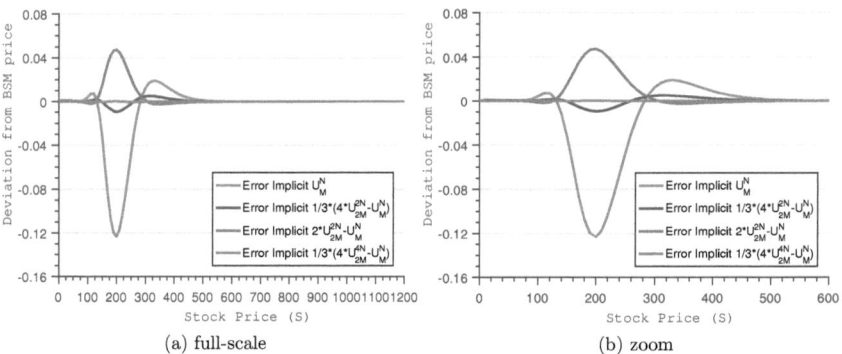

(a) full-scale   (b) zoom

Figure 4: Approximation Error for Implicit Scheme and Richardson Extrapolation

16

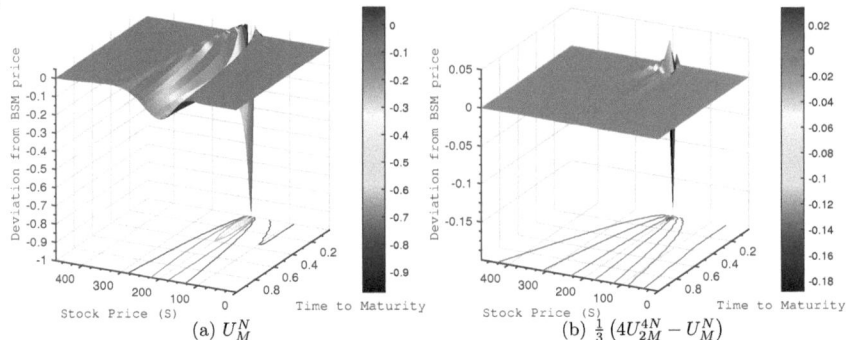

(a) U_M^N (b) $\frac{1}{3}\left(4U_{2M}^{4N} - U_M^N\right)$

Figure 5: Comparison of Approximation Error for Implicit Scheme

Coming back to the introductory example, in which it is shown that for the implicit scheme, an error reduction to 1 % is accompanied by 570 times more computational time. In contrast, the application of so-called *series acceleration methods*, like the Richardson extrapolation (RE), is able to reduce the error even to 0.22 % while computational time is only six times higher than before.

3.2 Richardson Extrapolation for Crank-Nicolson Scheme

For the CN scheme, the truncation error can be separated in a similar way in order to figure out efficient combinations of different grids:

$$\underbrace{h^2\left(\frac{a}{12}u_{xxxx} + \frac{b}{6}u_{xxx}\right)}_{=:v \text{ and independent of } h} + \underbrace{k^2\left(\frac{c}{8}u_{\tau\tau} - \frac{1}{24}u_{\tau\tau\tau}\right)}_{=:w \text{ and independent of } k} + \underbrace{k^2\left(\frac{a}{8}u_{xx\tau\tau} + \frac{b}{8}u_{x\tau\tau}\right)}_{=:z \text{ mixed term}}$$

In fact, the mixed term makes a similar derivation as for the implicit scheme quite tedious. Nevertheless, it can be shown that for similar combinations as for the implicit scheme the following numerical solutions can be achieved[13]:

$$1)\quad \frac{1}{3}\left(4U_{2M}^{2N} - U_M^N\right) = u(x,\tau) + O\left(h^4 + k^2h^2 + k^4\right)$$

$$2)\quad 2U_{2M}^{2N} - U_M^N = u(x,\tau) + O\left(h^2 + k^2h^2 + k^2\right)$$

$$3)\quad \frac{1}{3}\left(4U_{2M}^{4N} - U_M^N\right) = u(x,\tau) + O\left(h^4 + k^2h^2 + k^2\right)$$

Thus, one should stick to the first combination for an implementation of the CN scheme. Compared to an optimal Richardson Extrapolation for the implicit scheme, one only needs to double N which means relatively less additional computational effort. Furthermore, with this combination an even higher-order truncation error is obtained. Thus, an effective Richardson Extrapolation for the CN scheme results in a faster rate of convergence and features, in theory, a lower computational effort. Hence it is assumed to be superior.

---

[13] If requested, handwritten calculations can be submitted.

Table 2: Results for Richardson Extrapolation CN Scheme

	approx. error	comp. time	$\dfrac{\text{approx. error}}{\text{approx. error of } U_M^N}$	$\dfrac{\text{comp. time}}{\text{comp. time of } U_M^N}$
U_M^N	0.0936	0.0045	100 %	100 %
$\frac{1}{3}\left(4U_{2M}^{2N} - U_M^N\right)$	0.000172	0.0209	0.18 %	464.44 %
$2U_{2M}^{2N} - U_M^N$	0.0470	0.0202	50.21 %	448.89 %
$\frac{1}{3}\left(4U_{2M}^{4N} - U_M^N\right)$	0.000156	0.0356	0.167 %	791.11 %

However, a comparison of Table 1 and Table 2 shows that there is not necessarily a lower computational effort of the effective RE with a CN scheme compared to the fully implicit scheme. Nevertheless, the lower approximation error for the RE with the CN method is in line with the derived order of the truncation error. What is actually a little bit surprising is the difference in the approximation error for the combination 1 and 3. This is visualized in Figure 6. Actually, one should expect that the first combination yields more accurate results as it is of higher-order accuracy. Still, Figure 6 paints a quite different picture. The third combination is even a little bit more accurate, although the temporal order is lower. This issue is picked up in the Appendix again. For now, the first combination can be defined as the efficient RE due to the much lower computational time.

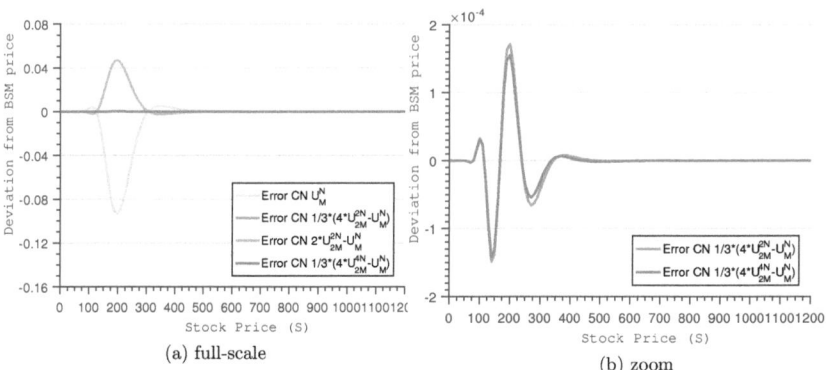

(a) full-scale

(b) zoom

Figure 6: Approximation Error for CN Scheme and Richardson Extrapolation

For this effective combination, the error development is contrasted to the error development of the simple CN scheme (Figure 7). As for the fully implicit scheme, this Figure demonstrates the fast rate of convergence that can be obtained with a series accelerating method.

In order to conclude this Section, Figure 8 compares the approximation errors for the implicit scheme, the CN scheme and their effective RE. As already mentioned above, the RE for the CN scheme is a little bit more accurate than the RE for the fully implicit scheme. However, it remains the question which of these two methods is more efficient with respect to the tradeoff between error reduction and additional computational effort. This issue is addressed in the next Section.

18

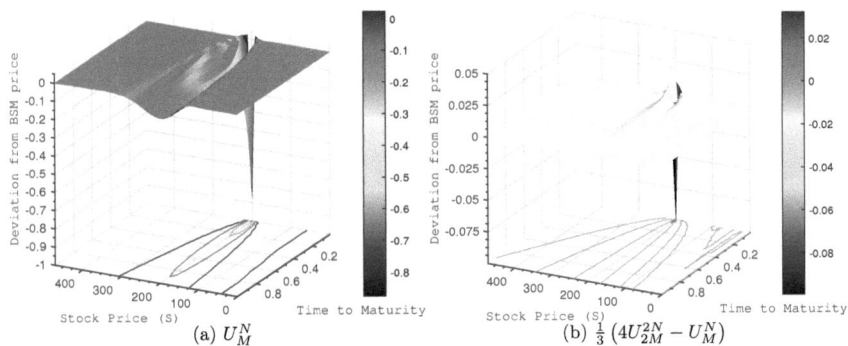

(a) U_M^N (b) $\frac{1}{3}\left(4U_{2M}^{2N} - U_M^N\right)$

Figure 7: Comparison of Approximation Error for CN Scheme

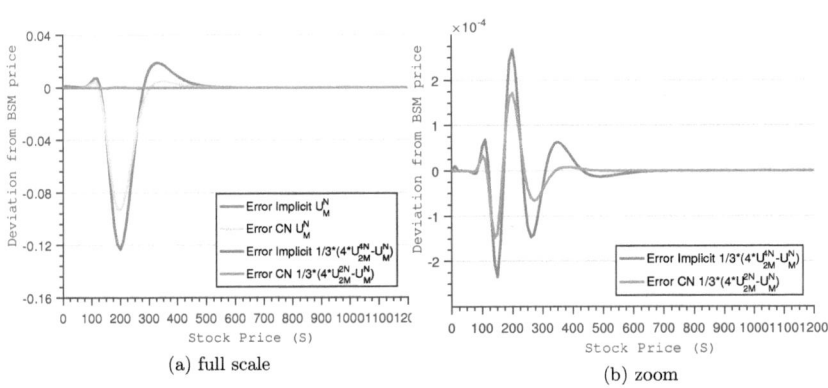

(a) full scale (b) zoom

Figure 8: Comparison of Approximation Error for both Schemes with RE

3.3 Preliminary Efficiency Analysis

In order to analyze how efficient the four methods are, the relation between N and M is fixed. As a first relation, $N = \frac{M^2}{72}$ for $M \in [12 : 12 : 1200]$ is considered. On the one hand, Figure 9a shows that both RE as well as the plain CN and fully implicit scheme yield approximately the same approximation error. The difference between the schemes with and without RE increase considerably when the number of total grid points is increased[14]. This might be explained by the higher leading error terms in the truncation error. On the other hand, Figure 9b clarifies the influence of the total number of grid points on the overall computational time. The results are in line with Table 1 and Table 2 from the previous Section.

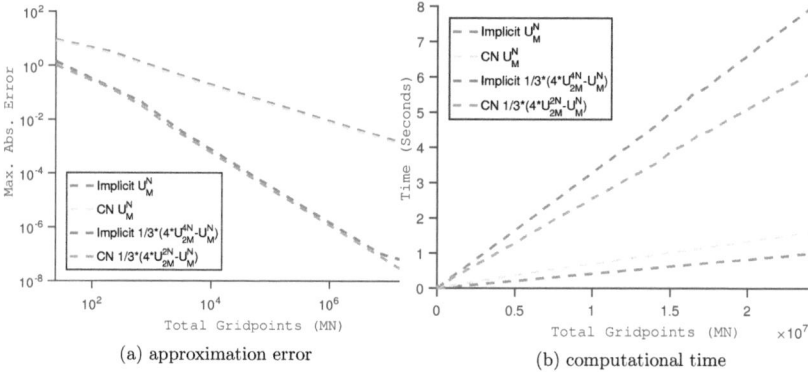

(a) approximation error (b) computational time

Figure 9: Influence of Total Number of grid points using relation $N = \frac{1}{72}M^2$

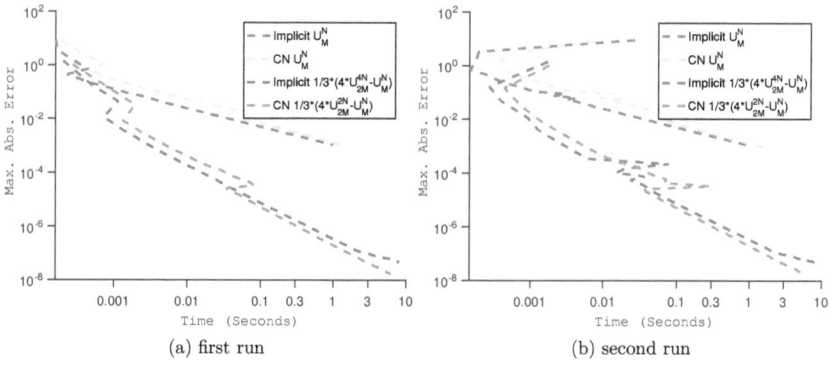

(a) first run (b) second run

Figure 10: Maximum Absolute Error as a Function of Computational Time with $N = \frac{1}{72}M^2$

For the tradeoff between computational time and approximation error, the superiority of the CN scheme with RE can be established when looking at Figure 10. The right plot is obtained from a second run as a robustness check. Both plots employ log-log scales meaning that up to

---

[14]Note that the plot employs a log log scale.

0.1 seconds, there are quite few observations when the relation is fixed in a quadratic manner. From this point on, the better tradeoff for the CN scheme with RE is illustrated by the lower line. Interestingly, the plain implicit scheme seems to have a better tradeoff than the plain CN scheme. However, the underlying fixed relation increases spatial as well as temporal grid points, meaning that there are two different effects that are difficult to disentangle. More accurate statements require a ceteris paribus analysis as it is conducted in the Appendix.

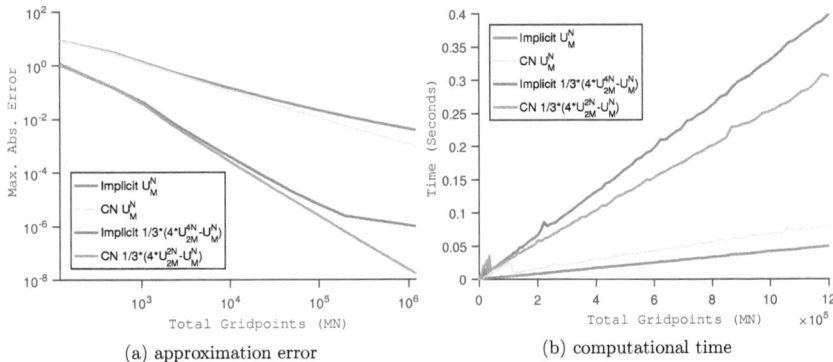

(a) approximation error

(b) computational time

Figure 11: Influence of Total Number of grid points using relation $N = \frac{1}{1.2}M$

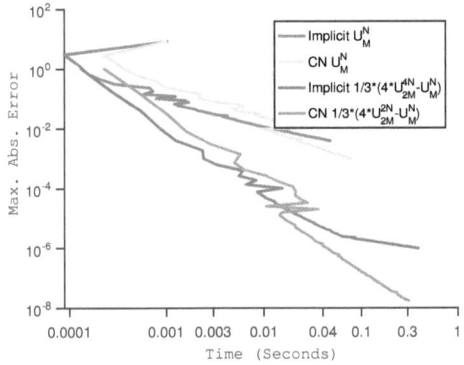

Figure 12: Maximum Absolute Error as a Function of Computational Time with $N = \frac{1}{1.2}M$

Nonetheless, some additional remarks can be made when using a second fixed relation according to $N = \frac{M}{1.2}$ for $M \in [12 : 12 : 1200]$. In Figure 11, the effect of the total number of grid points on the approximation error as well as the necessary computational time are put aside each other. No further conclusion can be drawn, except for the fact that the lines for the CN scheme and the fully implicit scheme seem to diverge for an increasing number of M. The respective tradeoff is illustrated in Figure 12. Due to the non-quadratic increase in N, this relation yields more observations with low computational effort. Considering this low computational area, the plain implicit scheme as well as the implicit scheme with RE seems to be more efficient compared to the respective CN specifications. However, after a certain point this advantageousness changes in favor of the CN method.

21

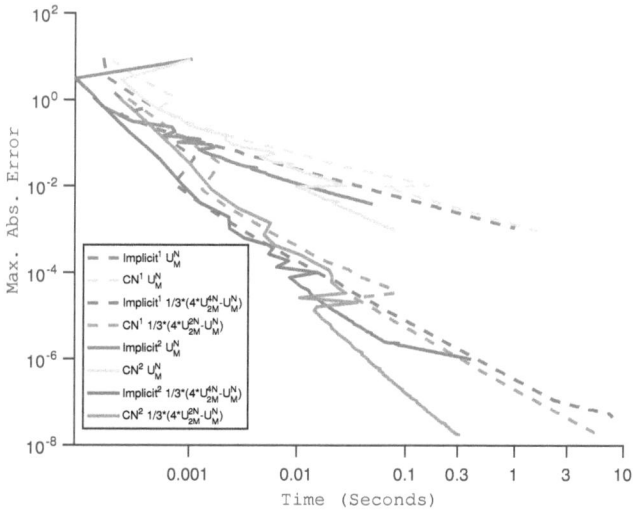

Figure 13: Comparison of Maximum Absolute Error as a Function of Computational Time

Finally, Figure 13 compares the tradeoff for both relations[15]. It can be concluded that a quadratic increase in N might be inefficient since most of the time the solid lines lie below the dotted lines. However, as already mentioned above, increasing grid points in space and time via a fixed relation simultaneously only yields some vague information concerning the tradeoff between accuracy and computational time. In order to show this, the obtained approximation errors of both relations are plotted against M and N (Figure 14). According to Figure 14a, one might assume that M is the main driver for error reduction and specifies the path for convergence, while a sufficiently large N is a sub-condition for this path of convergence. For the second relation, N is not increased sufficiently, such that the solid lines for the implicit scheme cannot catch up with the speed of convergence from the dotted lines. Turning to the CN scheme, another story can be told by looking Figure 14b. Increasing N further and further does not seem to have an impact on the performance of the CN scheme. This can be verified by the fact that for the CN scheme, the solid and the dotted line yield approximately the same error for a given M.

Due to the fact that this is a little bit hard to see in Figure 14b, another plot fixes M to 120 and uses the same number of N as before for both relations. The result is given in Figure 15. From this Figure there are primarily two things that can be noted. Firstly, an increase in N is like sunk costs in computational time for the CN scheme as no further time reduction can be obtained. Secondly, the error for the fully implicit scheme converges to the approximation error of the CN scheme when N is increased further and further. These examples give rise to the necessity of a ceteris paribus analysis for coming to a conclusion about the efficiency for the presented implicit methods. In fact, up to this point, only one main conclusion can be drawn. The application of *series acceleration methods* is sensible for finite difference methods and should be used to obtain quite accurate results with low computational effort.

---

[15]Supercript 1 determines the first relation and superscript 2 the second relation.

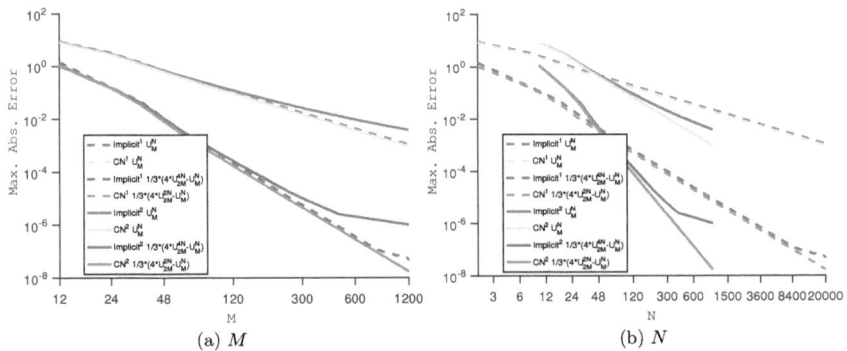

Figure 14: Comparison of the Effect of varying Grid Points on Approximation Error

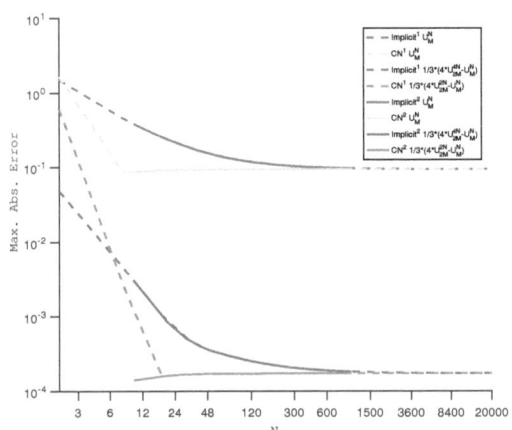

Figure 15: Comparison of Maximum Absolute for a fixed $M = 120$

3.4 Technical Implementation

The MATLAB - implementation of this Section is provided with *Task3.m*, *Task4.m* and *Task5.m*. For the calculation of the Richardson Extrapolations in Section 4.1 and 4.2, the functions *Implicit_put_RE.m* and *Implicit_put_CN.m* can be used. These functions are only designed to combine two different grids with each other. In order to do so, one needs to specify the respective grid points for both grids ($N1$, $N2$, $M1$ and $M2$) and additionally, how often these grids should be used for the extrapolation ($n1$ and $n2$). For these functions, the respective error developments are calculated in the usual manner.

As the error development comes along with additional computational time, modified functions for the efficiency analysis of Section 4.3 are designed which waive the calculation of the error development. These functions are called *Implicit_put_RE_mod.m* and *CN_put_RE_mod.m*, respectively. Only at the last time step, the respective deviation from the fair BSM price is calculated. For the combination of different grids, these functions also need to resort to modified functions for the implicit scheme as well as the CN scheme which also do not calculate the overall error development. These functions are called *Implicit_put2_mod.m* and CN_put4_mod.m and work as before, only waiving the error calculation.

Lastly, it needs to be mentioned that *Task4.m* simply loads a matrix with results which have been obtained before. This has to do with the run time of *Task4.m*. Due to the high number of N, the calculation of the Richardson extrapolation takes a lot of time. This inconvenience is bypassed by just loading the results into MATLAB. However, the code is commented out such that all necessary steps of calculations can be traced back. In general, more information about the MATLAB implementation can be found in the attached scripts.

Appendix

Spurious oscillations for Crank-Nicolson

In the theoretical session about the CN scheme, it was claimed that although the CN scheme is unconditionally stable, a phenomena called spurious oscillation can be observed. This becomes a problem if k is not small enough compared to h, or put differently, if there are relatively too less grid points in time. Hence, it is a bit comparable to stability issues of the explicit scheme. Nevertheless, there is no explosive behavior as it is the case when using an explicit scheme. Mathematically, spurious oscillation might be explained by a really slow convergence induced by a lot of eigenvalues pretty close to 1. The problem of spurious oscillation is illustrated in Figure 16 and Figure 17. As it is visible in Figure 16, the matter of spurious oscillation is no problem for an implicit scheme.

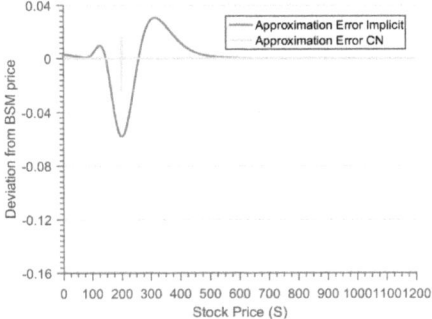

Figure 16: Spurious Oscillation of CN method for $M = 2000$ and $N = 50$

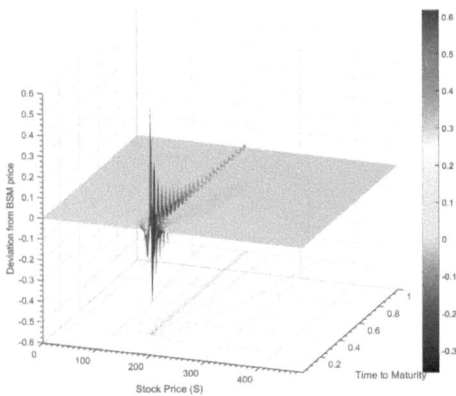

Figure 17: Spurious Oscillation of CN method for $M = 2000$ and $N = 50$ 3D plot

Problems of Crank-Nicolson for non-smooth Boundaries

In the main text, it is shown that there might be a little problem with the left-hand boundary in case of the fully implicit method. However, the CN scheme is also supposed to be quite sensitive to boundary conditions. This is especially true for non-smooth boundary conditions. In the numerical analysis above, a European put with a year to maturity is considered. If the time to maturity is increased further and further, a problem concerning the right-hand boundary can be observed. This is depicted in Figure 18 for $T = 5$ and $T = 10$. It seems that the assumption concerning a zero value of the option Delta is not appropriate when time to maturity is increased. An upgrade concerning a second derivative assumption, i.e. the Gamma, should be considered as this might mitigate this problem.

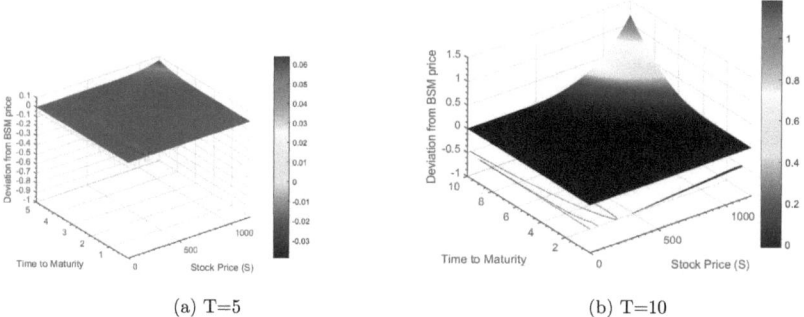

(a) T=5 (b) T=10

Figure 18: Righthand-Boundary Problem of CN method $M = 500$ and $N = 1000$ 3D plot

Continued Efficiency Analysis

This Section picks up the issues that were encountered during the efficiency analysis in Section 3.3. During my analysis, I try to figure out what the leading error term in the truncation error really means and how it affects the convergence. As a starting point, I have a look at the resulting approximation error for the fully implicit method and the CN method as well as their optimal REs when N is fixed. Figure 19 shows how the number of temporal grid points affects the approximation error for the same varying number of spatial grid points. There are several things than can be noticed[16].

Firstly, for both N, there seem to be critical numbers of spatial grid points up to which the implicit scheme and the CN scheme do not show different approximation errors. However, when this critical point is reached, the error reduction for the implicit scheme slows down while the CN scheme still converges at the same rate. Finally, the error reduction of the implicit scheme arrives at a bound or minimum error as additional grid points in time do not have a beneficial effect on the approximation error. Apparently, an increase in N shifts this critical point further to the right, i.e. more M are possible until the implicit scheme reaches this bound. Furthermore, the difference between the plain implicit scheme and the RE seems to be determined by the truncation error.

---

[16]These observations probably only hold for small T.

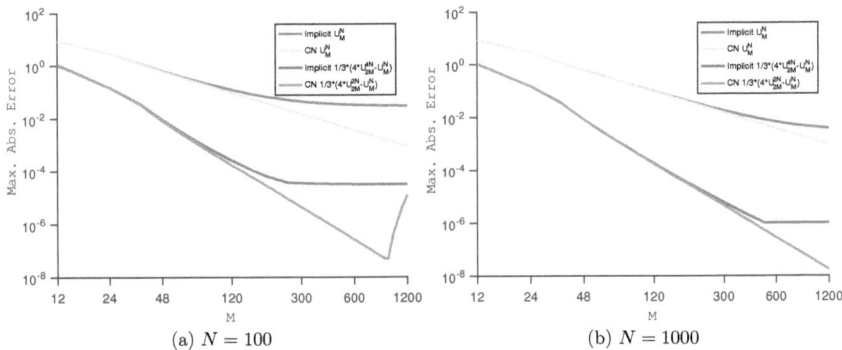

(a) $N = 100$ (b) $N = 1000$

Figure 19: Approximation error for fixed N and varying M

Secondly, considering the CN scheme, the number of time intervals is irrelevant for the absolute value of the approximation. With the exception of a small part of the orange line in the left plot, the lines for the CN scheme are exactly the same in both plots. The different behavior of the approximation error for high M when N is fixed to 100 can be attributed to the already mentioned spurious oscillation problem. As in case of the implicit scheme, the difference between the plain CN scheme and the optimal Richardson Extrapolation is determined by the leading error terms in the truncation error.

When Figure 19a is extended, i.e. M is increased up to 5,000 grid points, the argumentation from above is confirmed. Figure 20 shows that for a given number of grid points in time, the error of the CN scheme decreases further and further until it suddenly rises again and converges from below to the bound of the approximation error for the implicit scheme. Thus, as soon as spurious oscillations start to become a problem, the second-order accuracy with respect to time reduces apparently to first-order accuracy leading to the same bound as for the implicit scheme. For the RE implementation of the CN scheme, spurious oscillation is even a worse problematic feature. Instead of converging to the bound for the RE of the implicit scheme, it also converges to the bound for the plain implicit scheme. Actually, this Figure nicely depicts the different converging behaviors.

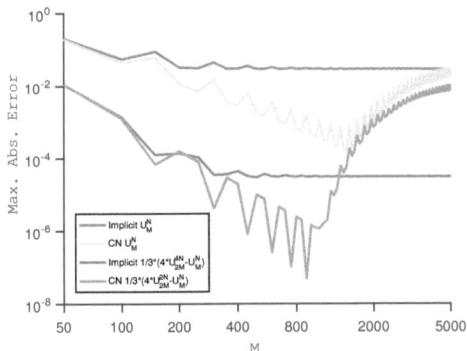

Figure 20: Convergence of the Approximation Error for $N = 100$ and varying M

Of course, a ceteris paribus analysis with respect to a varying N can also be conducted. The result is given in Figure 21. Looking at this plot, it seems that the approximation error using the fully implicit scheme converges in the limit, i.e. for sufficiently large N, against the approximation error of the CN scheme. Hence, for the same leading-error in space a lower bound for the approximation error is already determined. Due to a higher leading-error in time, the CN scheme reaches this bound much faster. An intersection of both lines is only possible when the problem of spurious oscillations prevent the CN scheme from converging with higher-order time accuracy. For instance, this is the case for both RE in the right plot. As small values of N evoke spurious oscillations for the RE of the CN scheme, the approximation error is at first much higher compared to the RE of the fully implicit scheme. After N has been sufficiently increased, it converges much faster to the predetermined lower bound and therefore "passes" the converging line for the fully implicit scheme with RE.

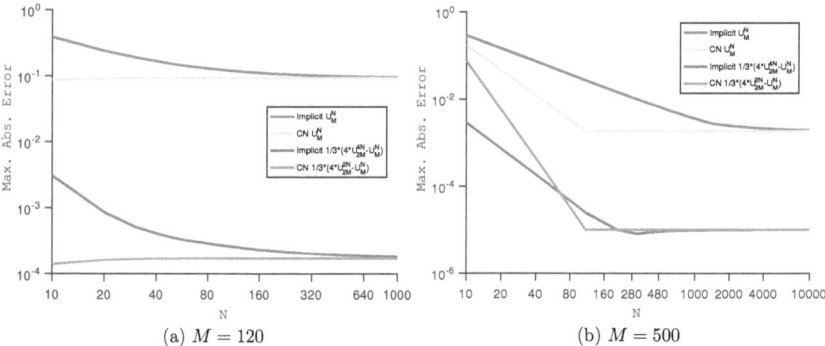

(a) $M = 120$ (b) $M = 500$

Figure 21: Approximation error for fixed M and varying N

In summary, it can be established that for the given data, in particular the relatively small time to maturity, the number of spatial grid points is the main driver for the **asymptotic** approximation error to which the finite difference methods converge. As shown in Figure 21a, this bound is likely to be defined by the highest leading error with respect to space. While the implicit scheme only converges to this bound when the number of temporal grid points is increased sufficiently, the CN scheme only needs more than a critical number of temporal grid point in order to avoid spurious oscillation. Moreover, the implicit scheme implicitly faces an additional bound if N is fixed such that increasing M does not yield a further reduction. For implicit methods, this bound seems to be determined by the leading error term with respect to time as can be verified by the comparison of the plain implicit scheme and the RE in Figure 20. Lastly, grid point positioning seems to play the same role for the CN scheme as for explicit methods which is illustrated by the oscillating decreasing approximation error. As mentioned by Tavella & Randall (2000), if the strike coincides with a spatial grid point, the approximation by differential quotients yields the poorest results. They recommend that the strike is positioned halfway between two spatial grid points.

Putting everything into a recommendation for action, the best performance can be obtained by choosing the effective Richardson extrapolation with the Crank-Nicolson scheme. For this method, the lowest possible approximation error is only determined by the number of spatial grid points. After this number has been chosen, one should choose the minimum number of temporal grid points that are needed in order to avoid the problem of spurious oscillation for the most efficient implementation.

References

BRANDIMARTE, PAOLO. 2006. *Numerical Methods in Finance and Economics - A MATLAB-Based Introduction.* 2nd edn. Hoboken NJ: John Wiley & Sons.

BRENNAN, MICHAEL, & SCHWARTZ, EDUARDO. 1978. Finite Difference Methods and Jump Processes and the Princing of Contingent Claims: A Synthesis. *Journal of Financial and Quantitative Analysis*, **13**, pp. 461–474.

CRANK, JOHN, & NICOLSON, PHYLLIS. 1947. A practical method for numerical evaluation of solutions of partial differential equations of the heat conduction type. *Proc. Camb. Phil. Soc.*, **43**, pp. 50–67.

FUSAI, GIANLUCA, & RONCORONI, ANDREA. 2008. *Implementing Models in Quantitative Finance: Methods and Cases.* 1st edn. Berlin Heidelberg New York: Springer-Verlag.

HIRSA, ALI. 2013. *Computational Methods in Finance.* Boca Raton FL: Chapman & Hall / CRC Press.

KWOK, YUE-KUEN. 2008. *Mathematical Models of Financial Derivatives.* 2nd edn. Hong Kong: Springer.

RICHARDSON, LEWIS FRY. 1911. The approximate arithmetical solution by finite differences of physical problems including differential equations, with an application to the stresses in a masonry dam. *Philosophical Transactions of the Royal Society A*, **210**, pp. 307–357.

TAVELLA, DOMINGO, & RANDALL, CURT. 2000. *Pricing Financial Instruments: The Finite Difference Method.* 1st edn. San Francisco: John Wiley & Sons.

WILMOTT, PAUL, HOWISON, SAM, & DEWYNNE, JEFF. 1995. *The Mathematics of Financial Derivatives.* New York: Cambridge University Press.